BUILDING BLOCKS OF CHEMISTRY

CHEMICAL ELEMENTS

Written by Cassie Meyer

Illustrated by Maxine Lee-Mackie

a Scott Fetzer company
Chicago

World Book, Inc.
180 North LaSalle Street
Suite 900
Chicago, Illinois 60601
USA

For information about other World Book publications, visit our website at **www.worldbook.com** or call **1-800-WORLDBK (967-5325).**

For information about sales to schools and libraries, call 1-800-975-3250 (United States), or 1-800-837-5365 (Canada).

© 2023 World Book, Inc. All rights reserved. This volume may not be reproduced in whole or in part in any form without prior written permission from the publisher.

WORLD BOOK and the GLOBE DEVICE are registered trademarks or trademarks of World Book, Inc.

Library of Congress Cataloging-in-Publication Data for this volume has been applied for.

Building Blocks of Chemistry
ISBN: 978-0-7166-4371-5 (set, hc.)

Chemical Elements
ISBN: 978-0-7166-4374-6 (hc.)

Also available as:
ISBN: 978-0-7166-4384-5 (e-book)

Printed in India by Thomson Press (India) Limited, Uttar Pradesh, India
1st printing June 2022

WORLD BOOK STAFF
Executive Committee
President: Geoff Broderick
Vice President, Editorial: Tom Evans
Vice President, Finance: Donald D. Keller
Vice President, Marketing: Jean Lin
Vice President, International: Eddy Kisman
Vice President, Technology: Jason Dole
Director, Human Resources: Bev Ecker

Editorial
Manager, New Content: Jeff De La Rosa
Associate Manager, New Product: Nicholas Kilzer
Sr. Content Creator: William D. Adams
Proofreader: Nathalie Strassheim

Graphics and Design
Sr. Visual Communications Designer: Melanie Bender
Sr. Web Designer/Digital Media Developer: Matt Carrington

Acknowledgments:
Writer: Cassie Meyer
Illustrator: Maxine Lee-Mackie/ The Bright Agency
Series Advisor: Marjorie Frank

TABLE OF CONTENTS

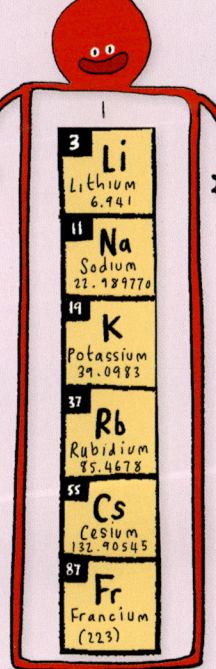

Introduction 4
What Are Chemical Elements? 8
Inside an Atom 10
The Birth of the Elements 12
Discovering and Classifying Elements...14
The Periodic Table 16
Element Symbols and Atomic Number.... 18
Atomic Mass 20
Periods 21
Groups 24
Classes of Elements 26
Classes of Metals 28
   Metalloids 33
   Nonmetals 35
   Conclusion 39
   Words to Know 40

There is a glossary on page 40.
Terms defined in the glossary
are in type **that looks like this**
on their first appearance.

INTRODUCTION

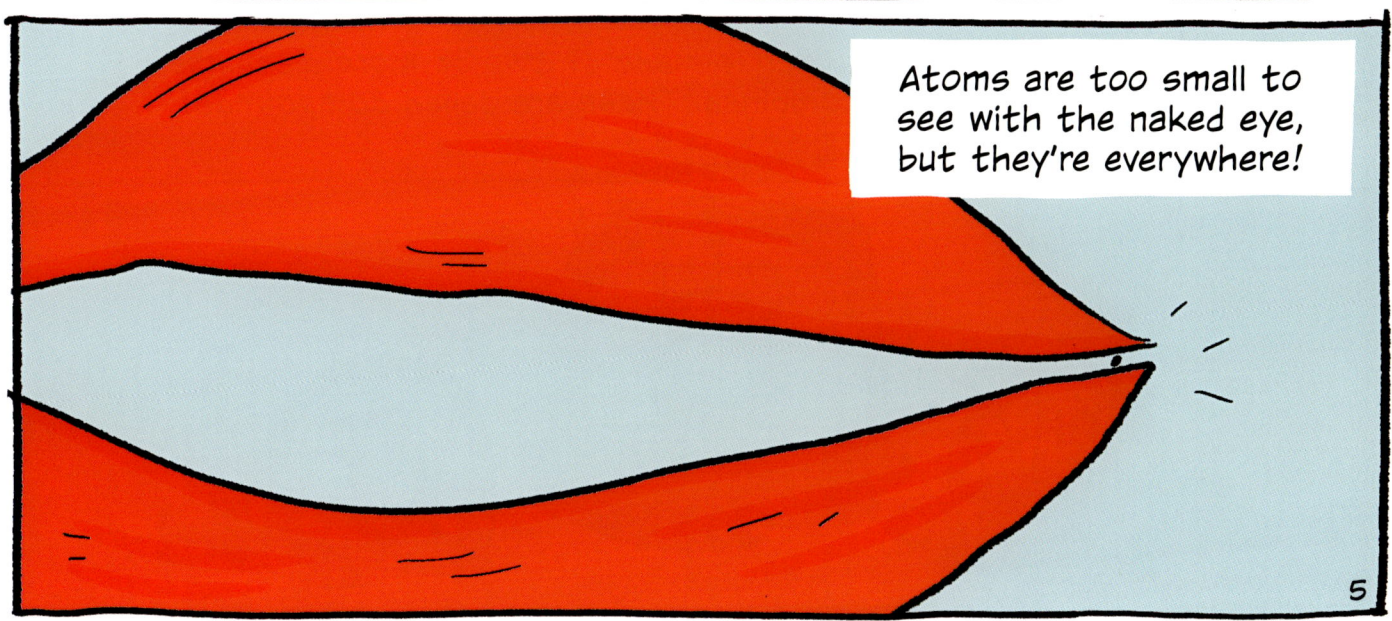

WHAT ARE CHEMICAL ELEMENTS?

There are 118 known chemical elements.

Of these, 93 are found naturally on Earth.

We're all-natural, man!

GROOVY!

COOL!

IT'S ALIVE!

CLANG

Scientists make the remaining elements in laboratories using special equipment.

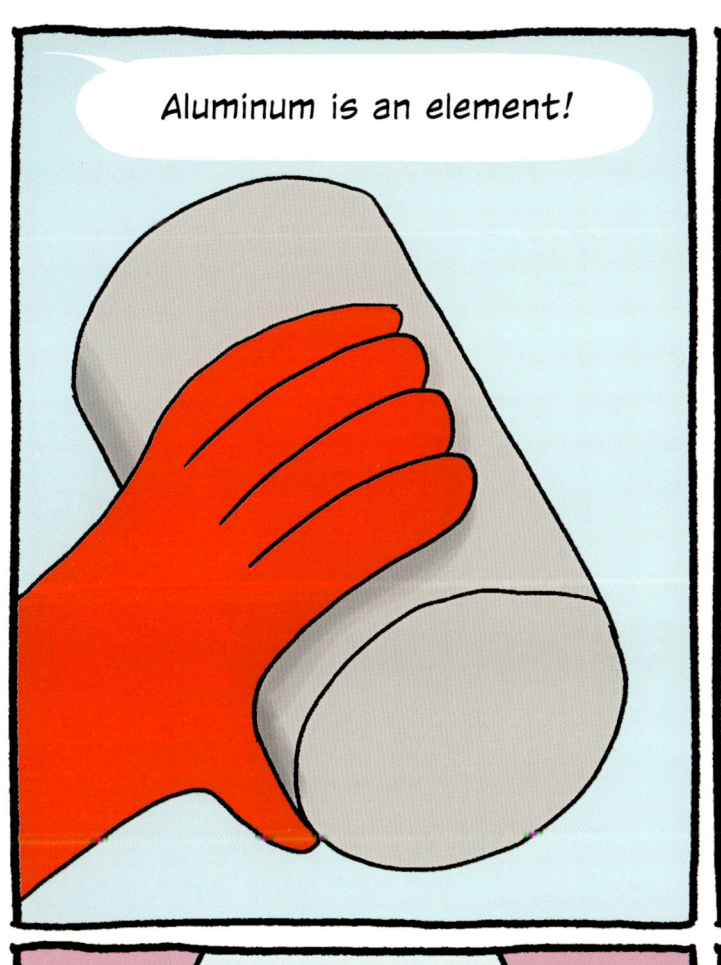
Aluminum is an element!

This lightweight metal is used to make everything from soft drink cans and cookware to buildings and airplanes.

Argon is an element! It is a gas at room temperature.

Some light bulbs are filled with argon gas to prevent corrosion (damage) around the metal filament.

9

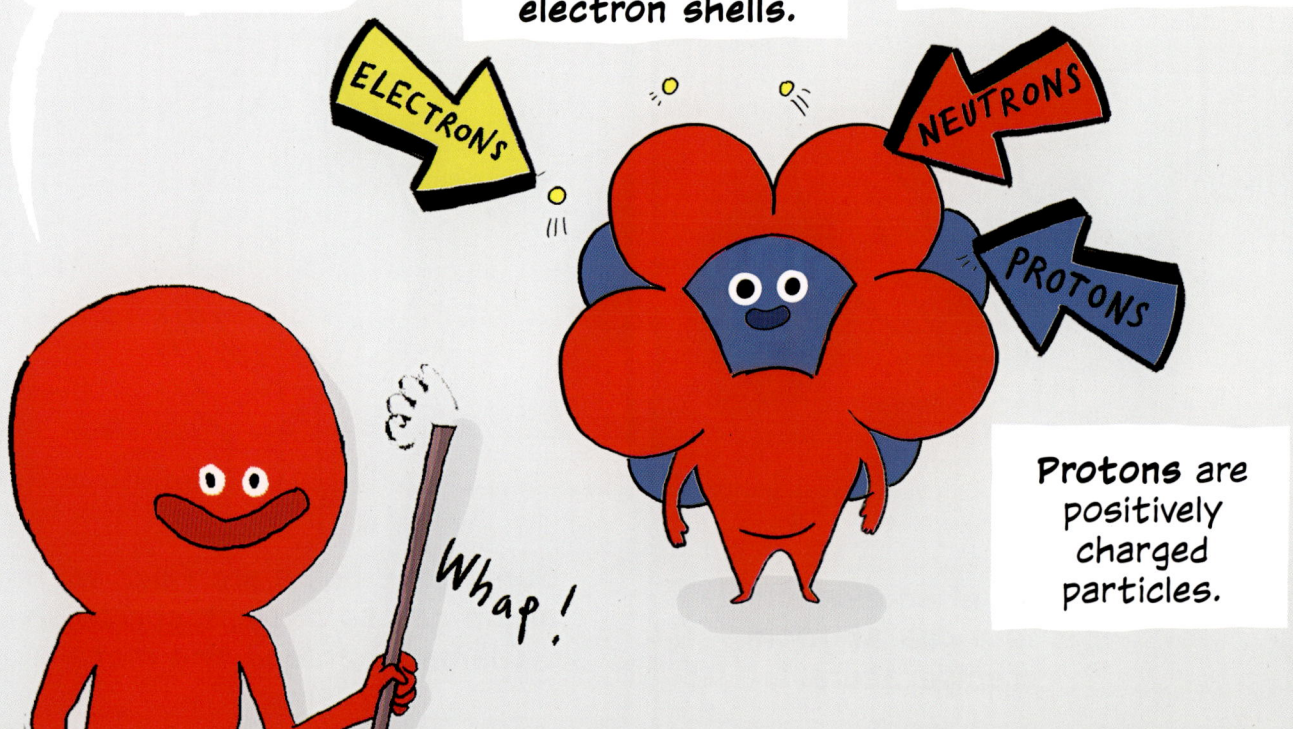

Most atoms of an element are identical, much like products from an assembly line.

But for some elements, a small percentage of atoms have a different number of neutrons.

A limited edition, you might say!

Each atomic form of an element is called an **isotope**. An element's isotopes all have the same number of protons. But different isotopes have different numbers of neutrons.

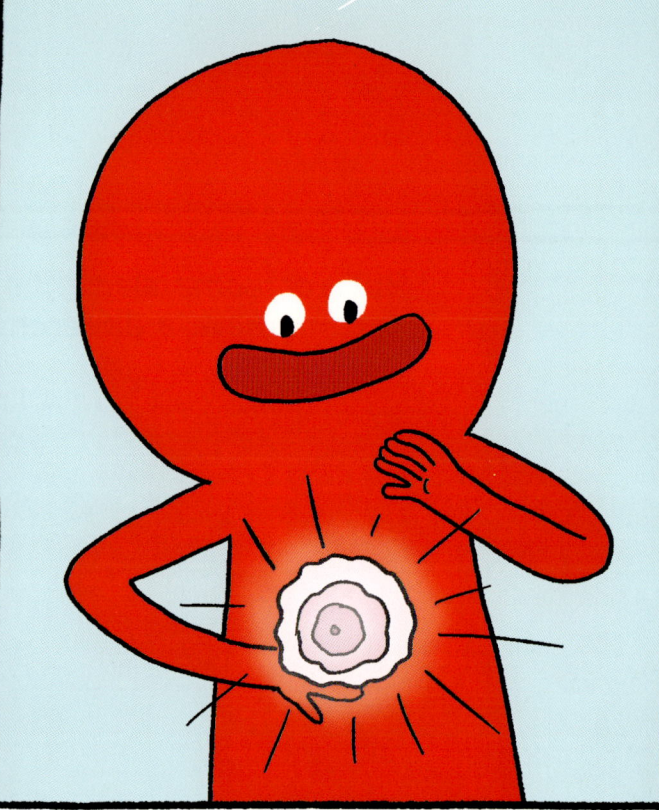

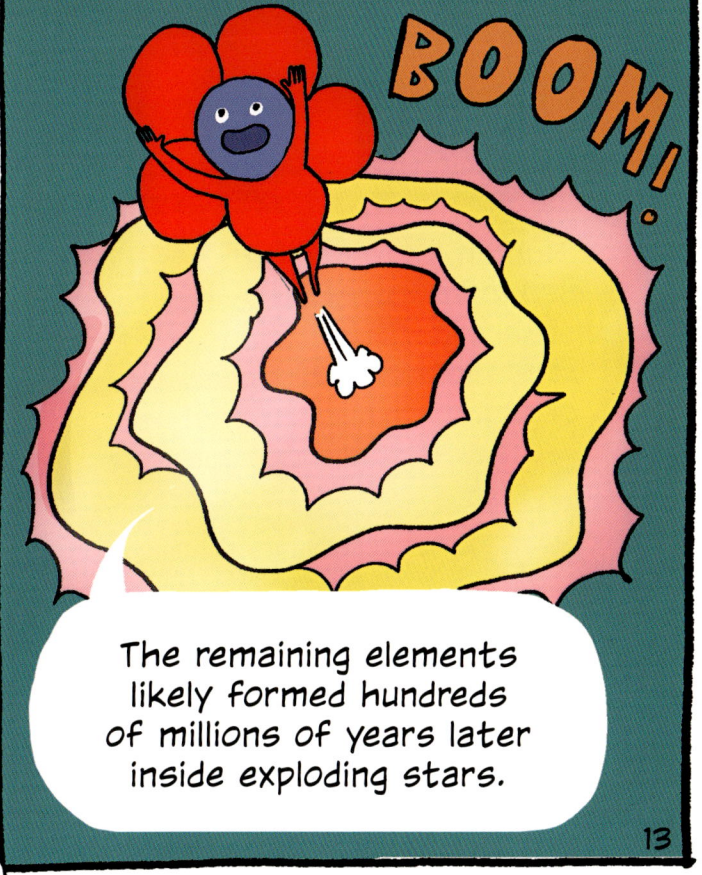

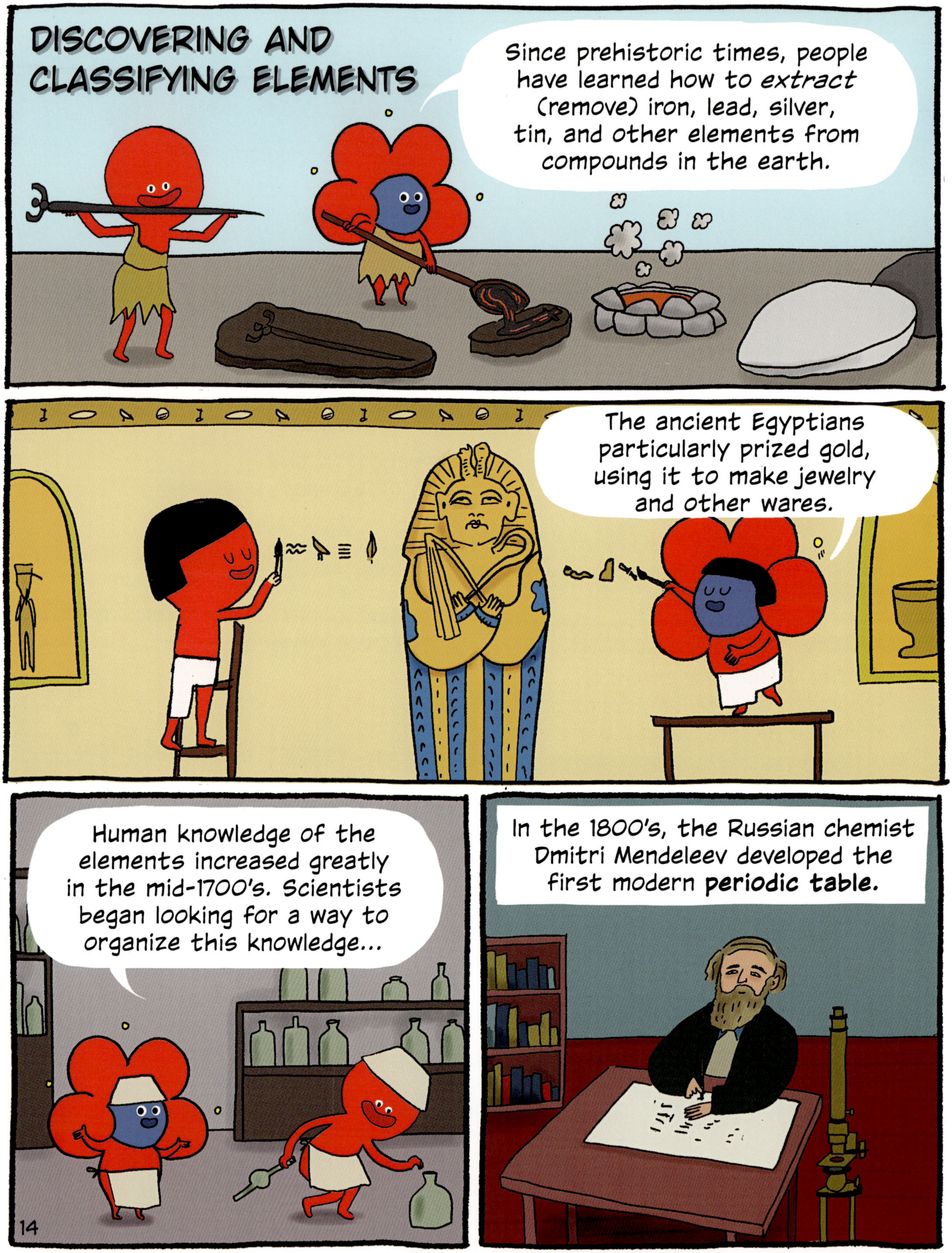

A periodic table shows the elements arranged according to their properties. Mendeleev's chart showed the known elements and even left gaps for elements yet to be discovered.

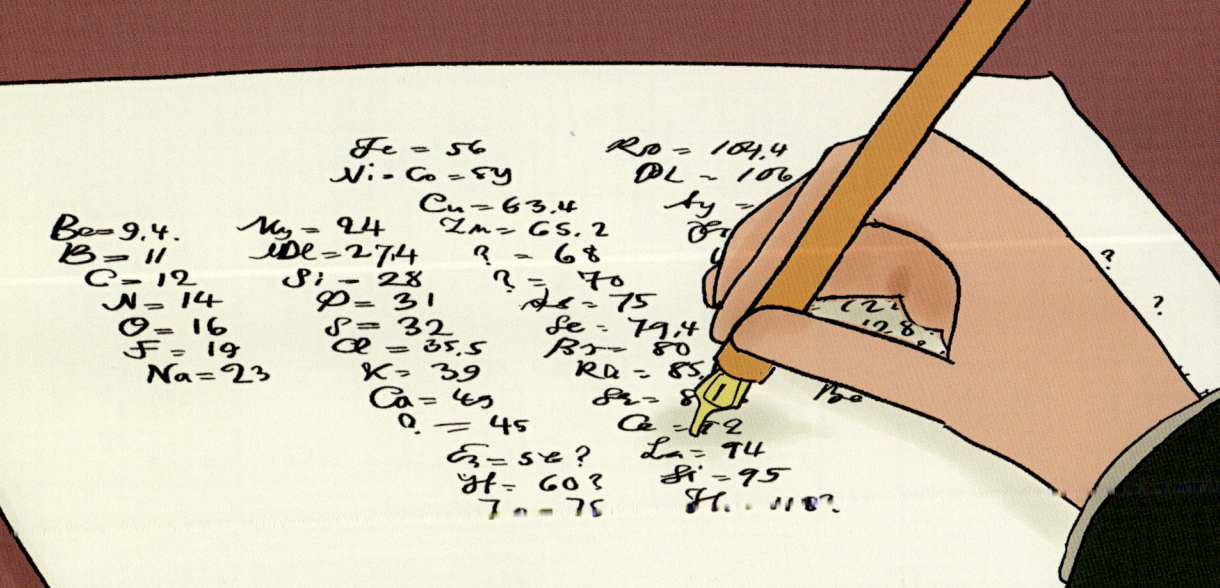

Over time, scientists modified the periodic table to include newly discovered elements.

Around the world, people continue to use the periodic table to understand the chemical elements.

THE PERIODIC TABLE

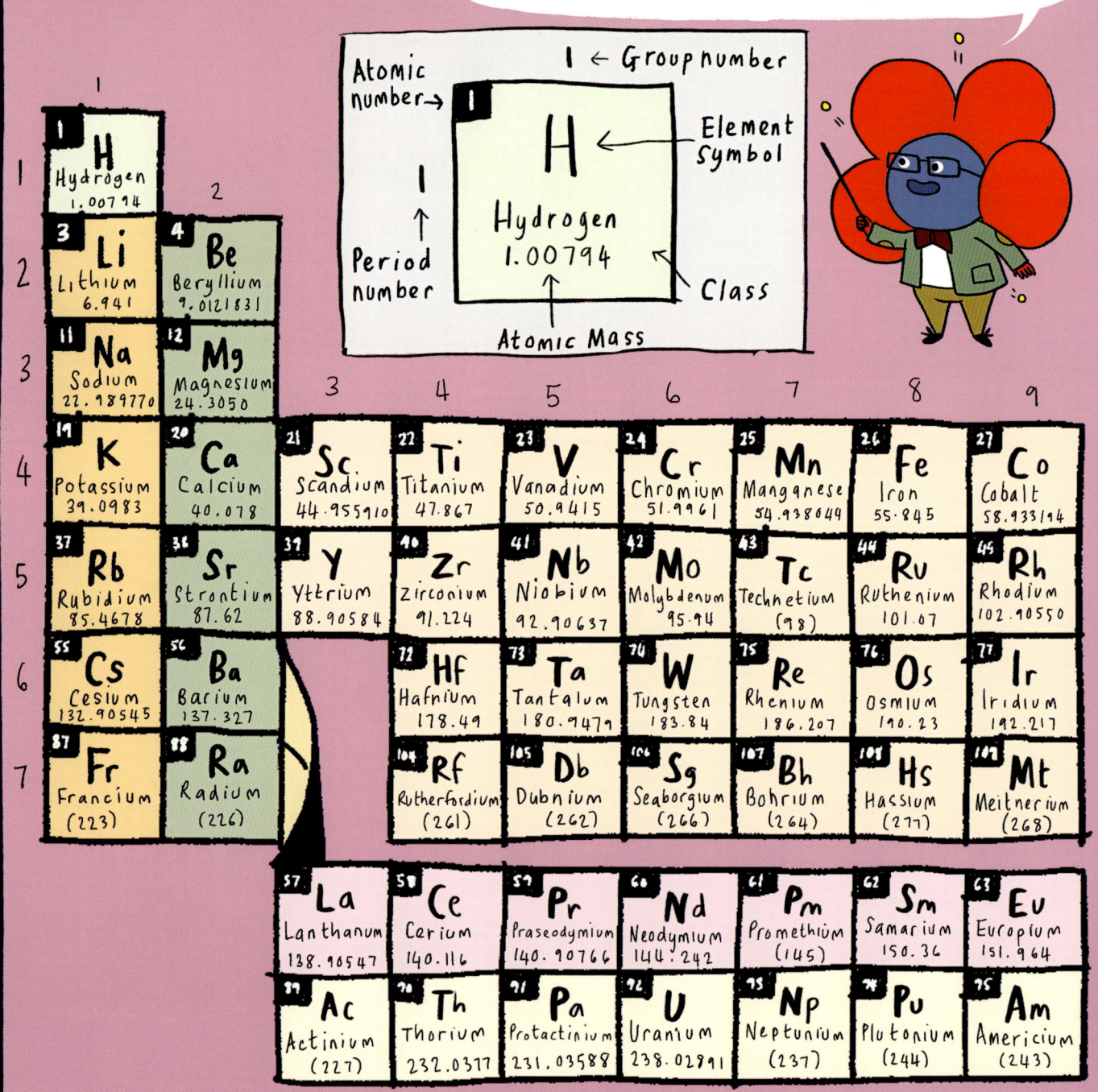

"The periodic table looks like it's written in code—and it is!"

"It includes symbols, numbers, letters, and even colors that reveal information about the properties of each element."

"Once you learn how to read the table, you can unlock this knowledge for yourself!"

Legend
- Alkali metals
- Alkaline earth metals
- Transition metals
- Lanthanide series (rare earths)
- Actinide series
- Other metals
- Metalloids
- Nonmetals
- Halogens
- Noble gases

Periodic Table (partial)

Group	13	14	15	16	17	18
						2 He Helium 4.002602
	5 B Boron 10.811	6 C Carbon 12.0107	7 N Nitrogen 14.0067	8 O Oxygen 15.9994	9 F Fluorine 18.998403	10 Ne Neon 20.1797
	13 Al Aluminum 26.981539	14 Si Silicon 28.0855	15 P Phosphorus 30.973762	16 S Sulfur 32.065	17 Cl Chlorine 35.453	18 Ar Argon 39.948

10	11	12	13	14	15	16	17	18
28 Ni Nickel 58.6934	29 Cu Copper 63.546	30 Zn Zinc 65.38	31 Ga Gallium 69.723	32 Ge Germanium 72.630	33 As Arsenic 74.921595	34 Se Selenium 78.96	35 Br Bromine 79.904	36 Kr Krypton 83.978
46 Pd Palladium 106.42	47 Ag Silver 107.8682	48 Cd Cadmium 112.414	49 In Indium 114.818	50 Sn Tin 118.710	51 Sb Antimony 121.760	52 Te Tellurium 127.60	53 I Iodine 126.90447	54 Xe Xenon 131.293
78 Pt Platinum 195.084	79 Au Gold 196.96657	80 Hg Mercury 200.592	81 Tl Thallium 204.3833	82 Pb Lead 207.2	83 Bi Bismuth 208.98040	84 Po Polonium (209)	85 At Astatine (210)	86 Rn Radon (222)
110 Ds Darmstadtium (281)	111 Rg Roentgenium (280)	112 Cn Copernicium (285)	113 Nh Nihonium (286)	114 Fl Flerovium (289)	115 Mc Muscovium (289)	116 Lv Livermorium (292)	117 Ts Tennessine (294)	118 Og Oganesson (294)

64 Gd Gadolinium 157.25	65 Tb Terbium 158.92534	66 Dy Dysprosium 162.500	67 Ho Holmium 164.93033	68 Er Erbium 167.259	69 Tm Thullium 168.93422	70 Yb Ytterbium 173.045	71 Lu Lutetium 174.9668
96 Cm Curium (247)	97 Bk Berkelium (247)	98 Cf Californium (251)	99 Es Einsteinium (252)	100 Fm Fermium (257)	101 Md Mendelevium (258)	102 No Nobelium (259)	103 Lr Lawrencium (266)

17

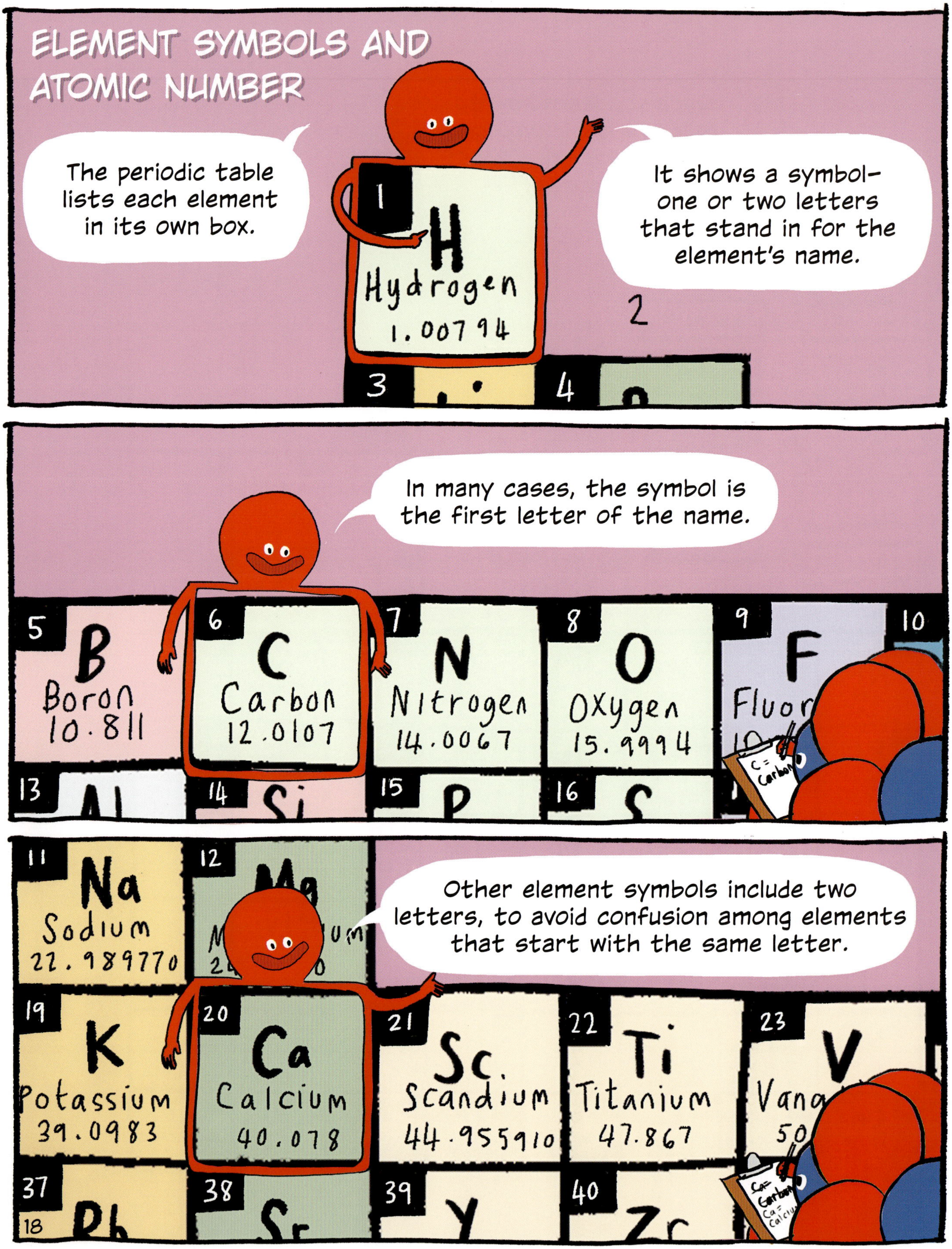

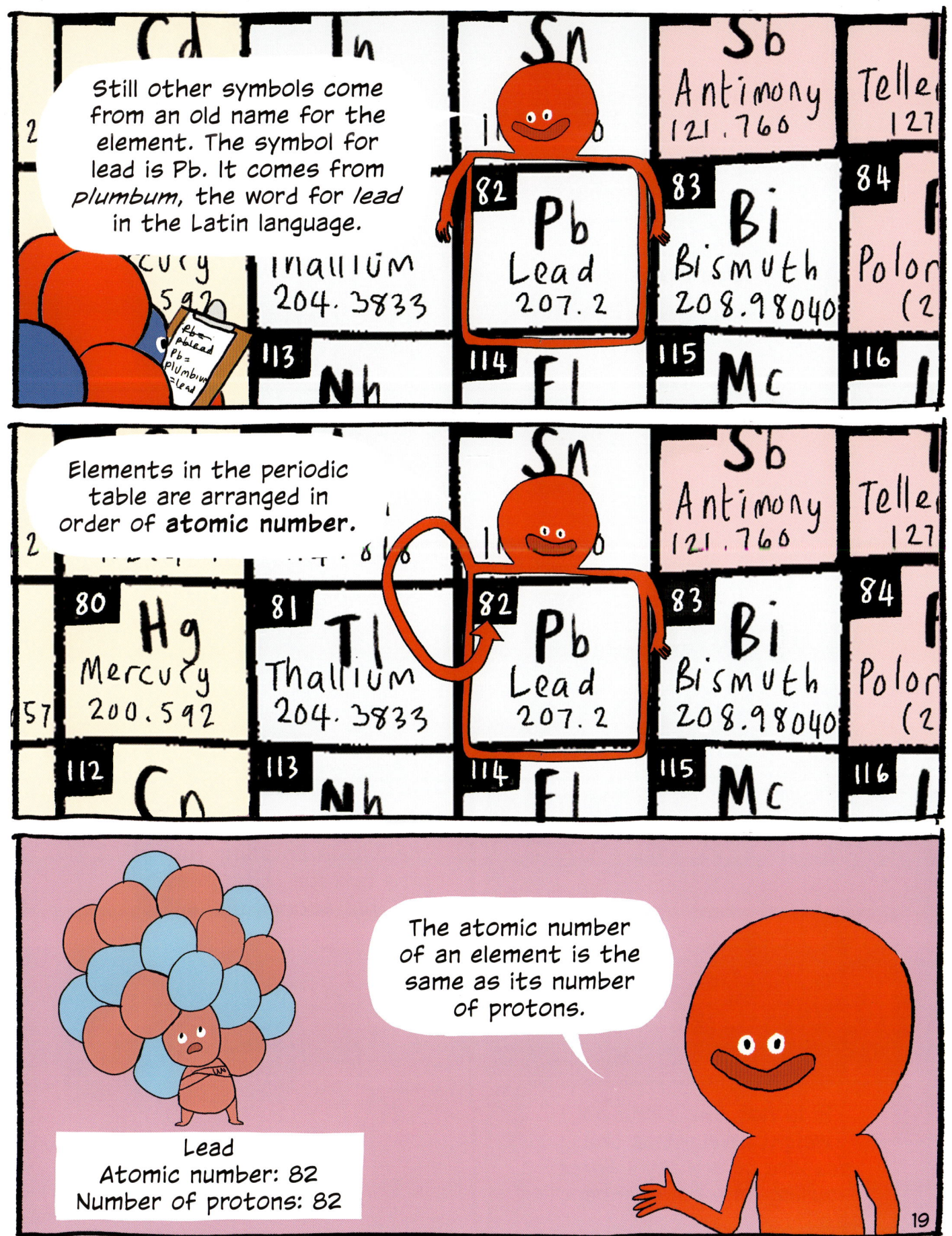

ATOMIC MASS

Some versions of the periodic table also show the **atomic mass** of each element.

Atomic mass is the amount of matter in the atom.

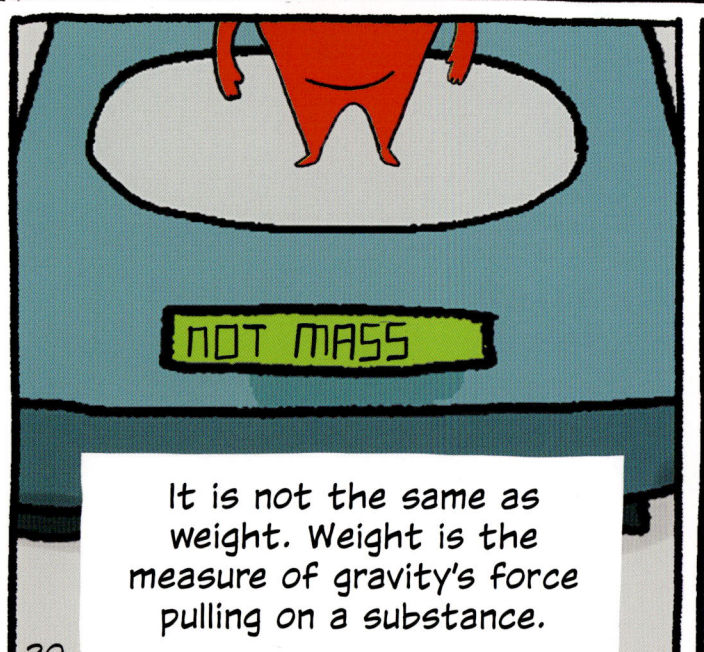

It is not the same as weight. Weight is the measure of gravity's force pulling on a substance.

The more protons and neutrons an atom has in its nucleus, the greater its atomic mass.

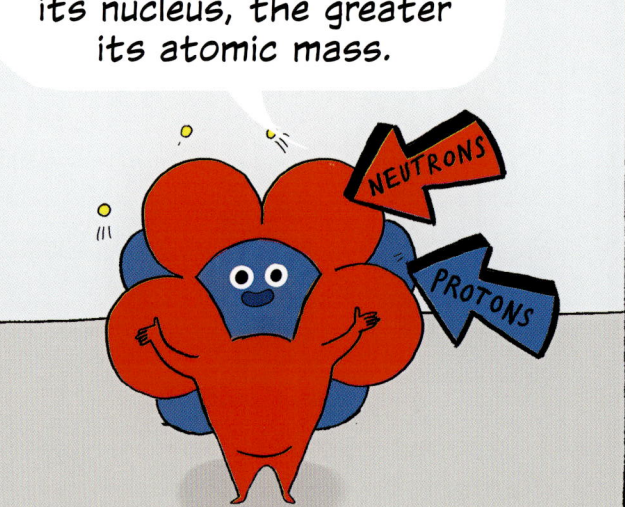

PERIODS

SEATING PLAN

Elements in the periodic table are listed in horizontal (side to side) rows, called periods. The periods aren't just thrown together—they have important meaning. Elements in each period have the same number of electron shells. An electron shell is like a particular layer of electrons around the nucleus.

The first period is a very exclusive row—just like the front row seats at a concert! It includes only two elements: hydrogen and helium. Atoms of these elements have only a single electron shell.

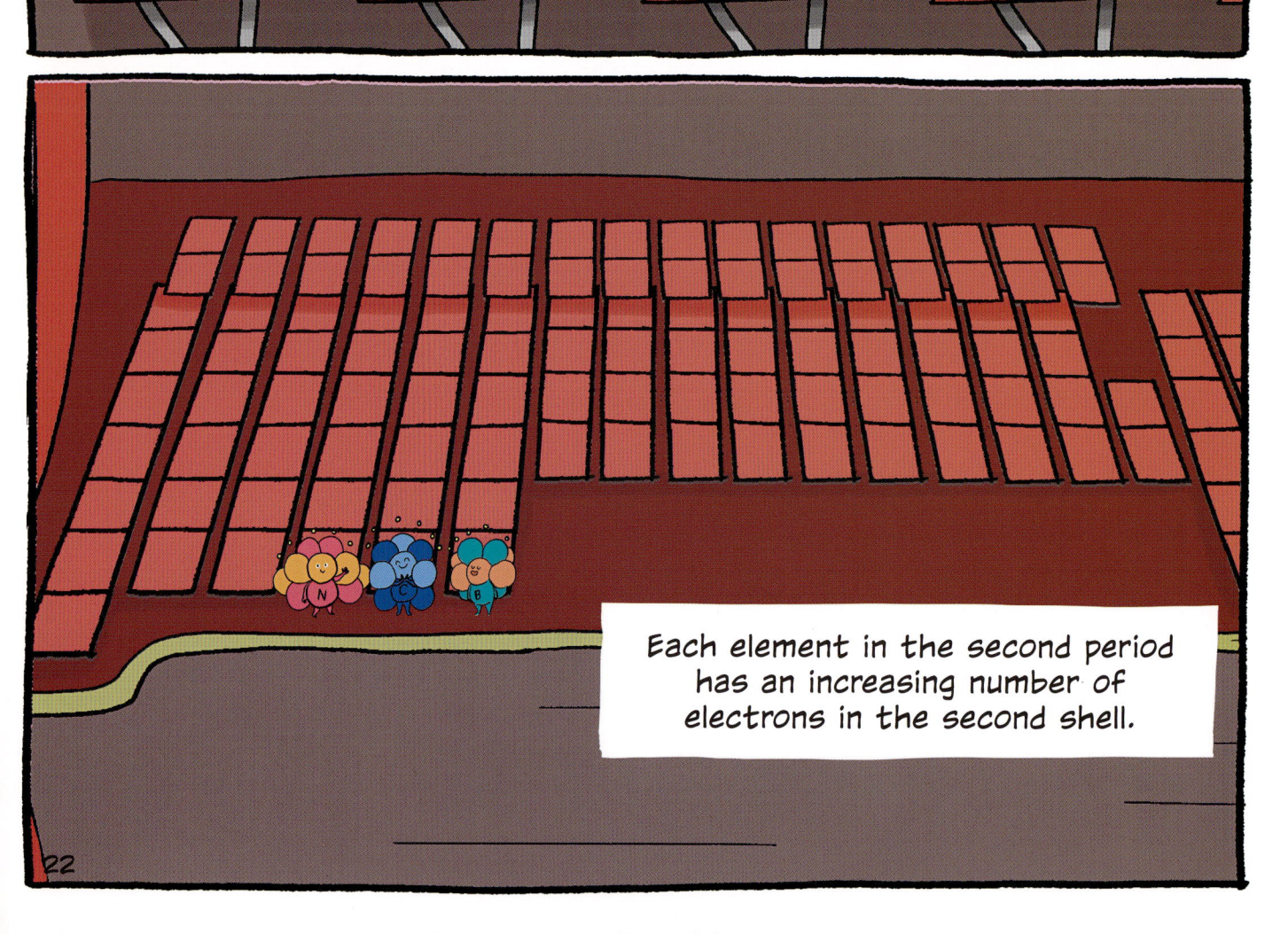

The element lithium begins the second period. Its atomic number is 3, and it usually has 3 electrons. The first shell can hold only 2 electrons, so it needs a second shell for number 3!

PERIOD 1   PERIOD 2   PERIOD 3   PERIOD 4

I'm so happy!

Each element in the second period has an increasing number of electrons in the second shell.

22

Remember, as the atomic number goes up, so does the number of electrons! Elements in the third period have three shells, and so on through the cheap seats in the seventh period.

ATOM SMASHER

Let's rock!

WOO! YAY!

23

GROUPS

Elements that have similar properties appear in columns called groups, or families. Each group has a number from 1 to 18.

Elements in a group form compounds in a similar way. This is sort of like the way traits are passed among family members.

Consider chemical reactions, for example. In a **chemical reaction,** an atom may gain, lose, or share electrons.

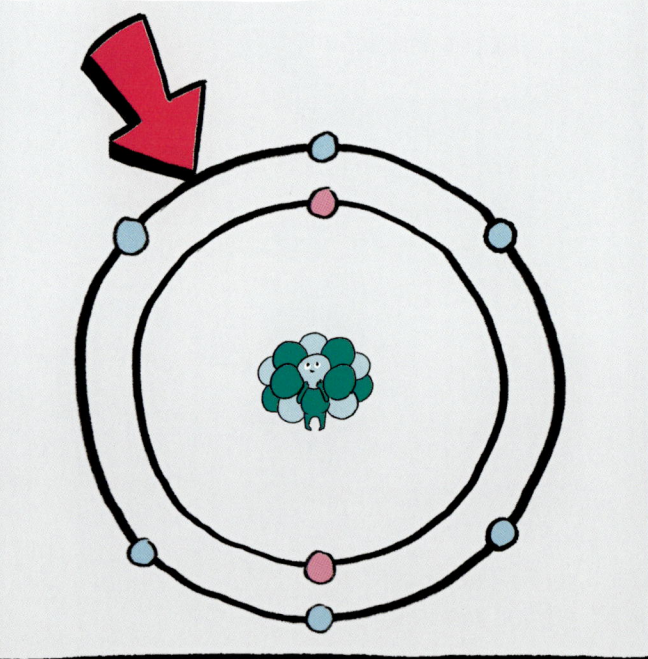

Only electrons in the outermost shell—called the **valence** shell—can be lost, gained, or shared. These are called valence electrons!

Elements in a group typically have the same number of valence electrons, so they form bonds with other atoms in the same way.

Groups on the left side of the periodic table typically lose electrons, gaining a positive electric charge. Groups on the right side tend to gain electrons, gaining a negative electric charge.

Lithium, for example, is a group 1 element. It can give one valence electron to fluorine to make the compound lithium fluoride. Most elements in group 1 bond this way.

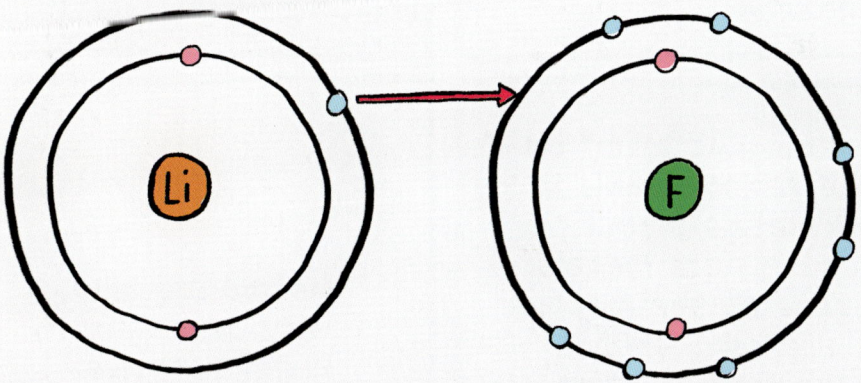

Chlorine is a group 17 element. It can share an electron from its valence shell with hydrogen—to form the compound hydrogen chloride.

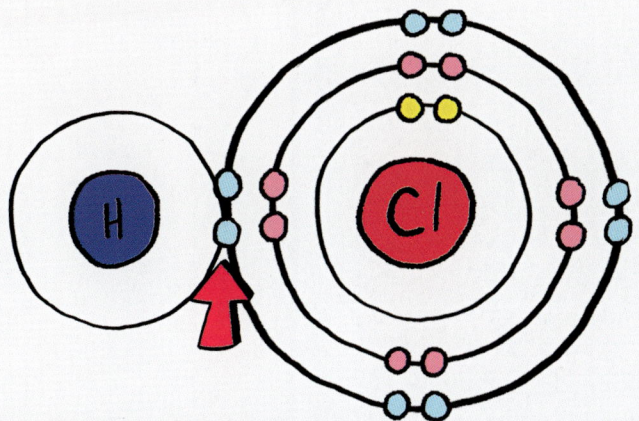

CLASSES OF ELEMENTS

An element can share similarities with other nearby elements in the periodic table—sometimes even from different periods or groups!

Think about a club: you might be in a club with family members, neighbors, or others who live farther away.

The biggest "club" on the periodic table is the metals. About 95 elements are metals. In general, the metals are hard, shiny, and solid at room temperature.

Metals are also very dense and highly reactive.

They are good **conductors** of electricity and heat.

26

And, metals can be easily shaped and even spun into thin wires—another quality that makes them useful for carrying electricity!

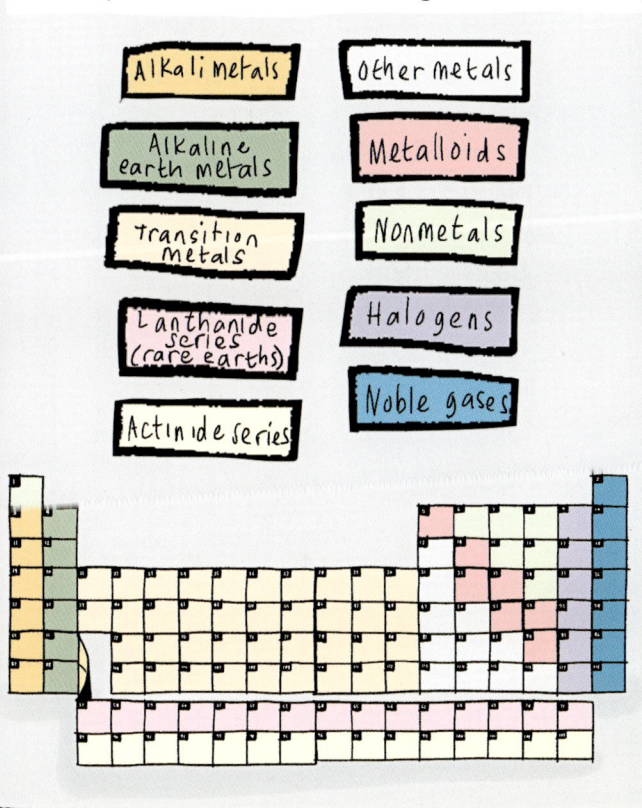

But not all of the clubs are as large as the metals. Each element belongs to one of 10 different smaller clubs called classes. The classes are generally shown on the periodic table using color.

Alkali metals
Alkaline earth metals
Transition metals
Lanthanide series (rare earths)
Actinide series
Other metals
Metalloids
Nonmetals
Halogens
Noble gases

Elements in a class share certain physical and chemical properties.

What class are you a part of, Atom?

I wouldn't want to belong to any class that would have me as a member.

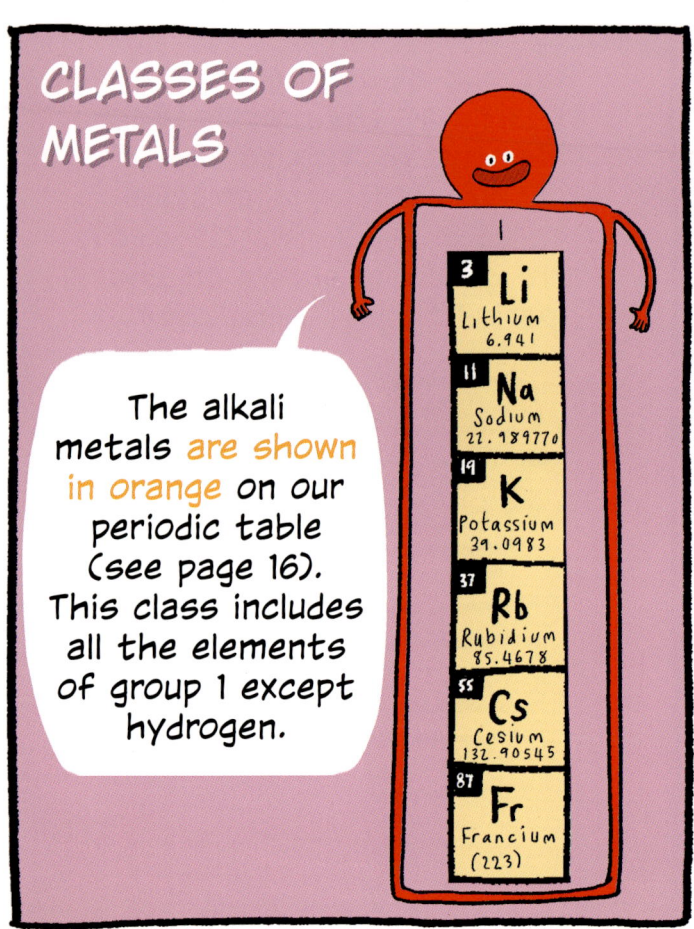

Most alkaline earth metals react with water, though less vigorously than alkali metals. Like alkali metals, they only appear in nature as compounds.

Some alkaline earth metals are used in electronics or X-ray technology. Others are used in making fireworks. Some are found in shells and bones. Calcium is even found in cheese.

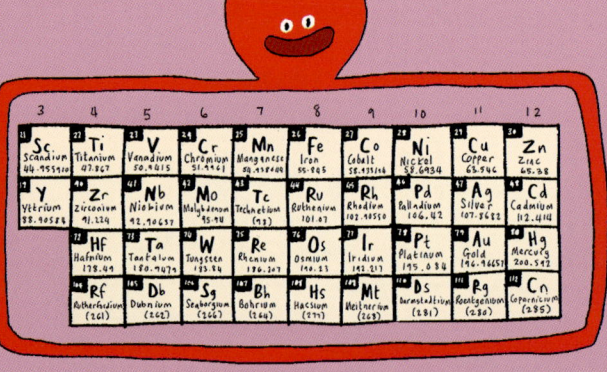

Transition metals are shown in light orange. Almost all the elements from groups 3 through 12 are transition metals. They include some of the most well-known metals, such as copper, gold, iron, nickel, and silver.

These metals are very hard and not reactive. They're often used to make things that are strong and won't react with their environment.

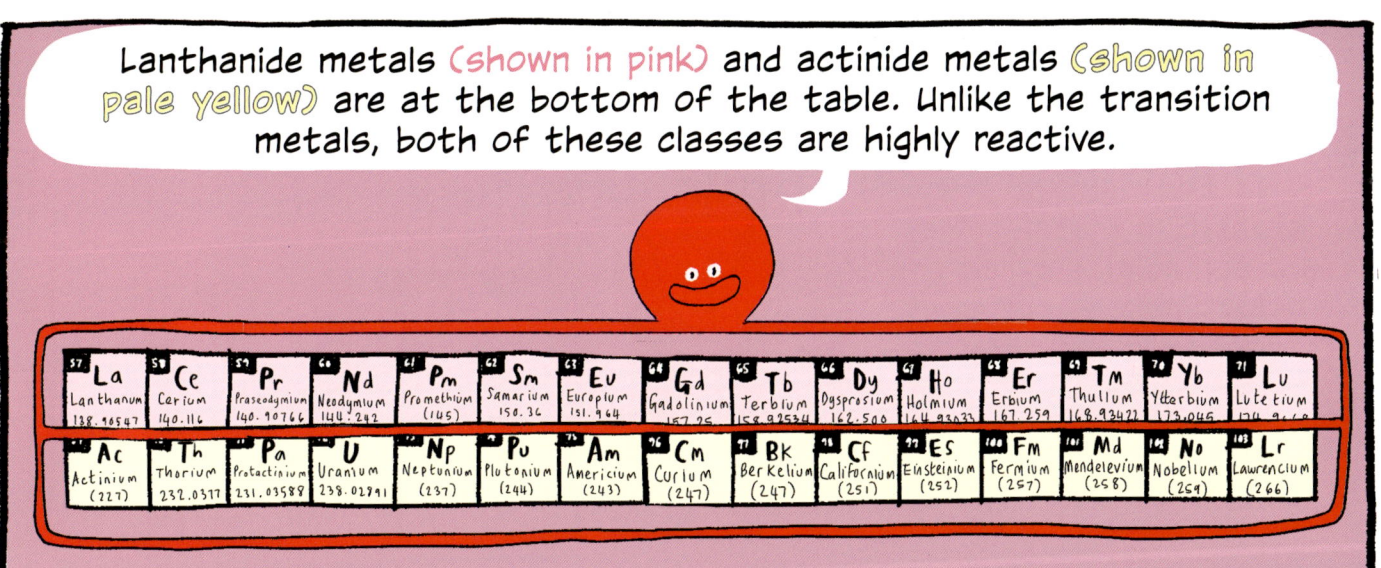

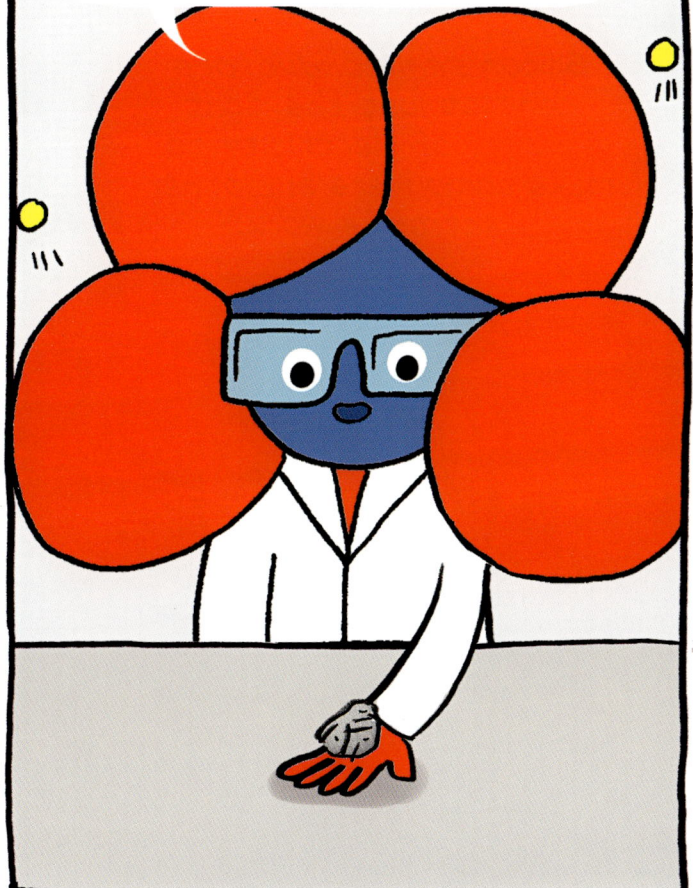

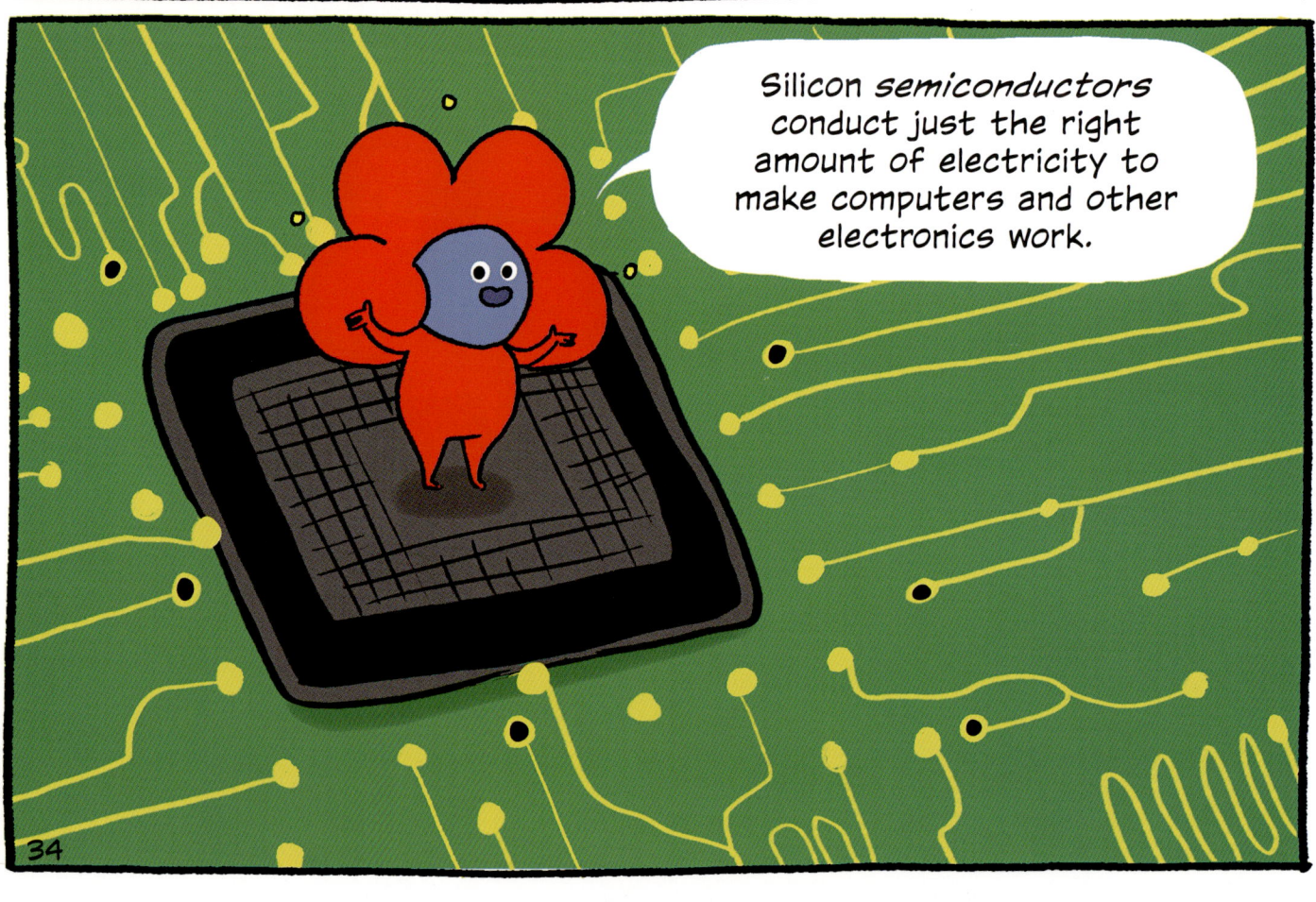

Electronics—which can be found in computers, smartphones, and pretty much everything else these days—need electricity of course, but only small amounts and in certain places.

SMARTCOUCH X

Silicon *semiconductors* conduct just the right amount of electricity to make computers and other electronics work.

NONMETALS

Not every element is a metal, of course. Elements that aren't metals, or *nonmetals*, have a few things in common.

Nonmetals are less dense than metals.

Solid nonmetals are brittle, dull, and cannot be shaped easily.

Nonmetals are not good conductors of heat or electricity, so they're often used as insulators for wires.

35

Nonmetal elements form different classes, too! Noble gases, shown with the color teal on our periodic table, are a separate class of nonmetals. They make up group 18.

These gases are not very reactive.

Off with his head!

Hi

Chemists gave these gases the name "noble" because they acted like kings, queens, and other nobles, refusing to bond with "lower classes" of atoms.

The lack of reactivity makes them very useful. Helium, for example, is lighter than air and does not burn. So, it's used to fill balloons and other airships.

The halogens are a highly reactive class of nonmetals. They're shown in the color lavender. All halogens hail from group 17.

Different halogens are solid, liquid, or gas at room temperature. They are yellow, green, red, brown, and purple in color.

Halogens tend to combine with metals to produce compounds called salts, including table salt.

Halogens are also found in toothpaste, photographic film, medical disinfectants, and swimming pools.

Take my picture by the pool, 'cause I'm the next big thing!

CONCLUSION

The periodic table ranks as one of the most useful tools in all of science.

As you learn more and more about it, you'll better understand the chemical elements.

You may even begin to appreciate the roles they play in the world around you.

There are so many treasures yet to be discovered in the world of chemistry.

Hard tellin' what you might find!

WORDS TO KNOW

atom one of the basic units of matter.

atomic mass the amount of matter in an atom.

atomic number the number of protons in the nucleus (center) of an atom of a chemical element. The atomic number is used in describing an element. All atoms of the same chemical element have the same number of protons.

chemical element a substance made of only one kind of atom. There are 118 chemical elements.

chemical reaction a process by which one or more substances are chemically converted into one or more different substances.

compound a substance that contains more than one kind of atom.

conductor a material that easily transmits heat, electricity, light, sound or another form of energy.

electron a kind of particle that circles around the nucleus (center) of an atom. Electrons have a negative electrical charge.

electron shell a grouping of electrons arranged at similar distances from the nucleus (center) of an atom according to how much energy they have.

isotope one of two or more atoms of the same chemical element that differ in the amount of neutrons they contain. Different isotopes of a chemical element will have different atomic masses.

neutron a kind of particle inside the nucleus (center) of an atom. Neutrons have no electrical charge.

nucleus the center of an atom. The nucleus is made up of protons and neutrons.

periodic table a chart that lists the known chemical elements arranged according to their characteristics.

proton a kind of particle inside the nucleus (center) of an atom. Protons have a positive electrical charge.

reactive describes a substance that readily combines with other substances.

valence a number that indicates the number of electrons available in an atom to form chemical bonds with other atoms.